Motion

A World on the Move

Glen Phelan

Sally Ride, Ph.D., President and Chief Executive Officer;
Tam O'Shaughnessy, Chief Operating Officer and
Executive Vice President; Margaret King, Editor;
Monnee Tong, Design and Picture Editor; Erin Hunter,
Science Illustrator; Brenda Wilson, Editorial Consultant;
Matt McArdle, Editorial Researcher

Program Developer, Kate Boehm Jerome
Program Design, Steve Curtis Design Inc.
www.SCDchicago.com

Sally Ride Science
9191 Towne Centre Drive
Suite L101
San Diego, CA 92122

ISBN: 978-1-933798-68-4

Printed in the United States of America
10 9 8 7 6 5 4 3 2
First Edition

Cover: Bicycle racers speed around a track. How would
you describe their motion?

Title page: Rhythmic gymnastics requires balance, skill,
and an understanding of motion.

Right: This team of the Navy's Blue Angels takes motion to
new heights.

Sally Ride Science is committed to minimizing its environmental impact by using
ecologically sound practices. Let's all do our part to create a healthier planet.

This book is printed on paper made with 100% recycled fiber, 50% post-consumer
waste, bleached without chlorine, and manufactured using 100% renewable energy.

Contents

Introduction

In Your World

Don't move! Are you sitting perfectly still? You may think you are—but you are moving more than you think!

The planet you're sitting on is zooming around the Sun. And there's lots of movement going on in and around you that you can't see. Blood rushes through your veins. **Molecules** of air zip past you as wind.

Of course, there's also a lot of movement going on that you *can* see. Some of it is fast. A dog chases after a squirrel. A rocket races into space. Lightning streaks across the sky.

Other things move slowly. Honey drips from a spoon. A sloth creeps along a tree branch.

The amazing thing about all this **motion** is that it doesn't matter if the moving object is large or little, fast or slow, visible or invisible. All moving objects—from rocket ships to sloths—follow the same rules.

Chapter 1: A Change in Position

Where, Oh Where?

"Ugh!" You lunge at the ball to return the serve, but it zips past your outstretched racquet in a yellow blur.

"Game, set, match!" your opponent, and friend, announces.

"I'll get you next time," you say between breaths. "Let's take a break. We can leave our stuff here." But how would you describe where you left your stuff?

When you talk about where an object is, you describe its **position**. How do you describe position? One way is to use something that doesn't move as a **point of reference**.

▲ **How would you describe the position of the tennis equipment?**

In this case, your point of reference might be the tennis net. You could say you left your stuff by the net. The net stays put, so it's a good point of reference. If you stepped away to get a drink of water, you'd know exactly where you could find your stuff, unless . . .

. . . You come back to find your stuff missing! You know where you left it because your point of reference—the net—is still there. Your stuff, however, has changed position. Rather, someone changed your stuff's position.

About the Position . . .

People use a point of reference nearly every time they talk about something's position.

You might say, "The fireworks are down by the bay tonight." Or, "The blimp is over the stadium." Each description uses a point of reference.

You can even describe the same object using different points of reference. For example, think about how you might describe where the dog is in the picture on this page. If you use the fire hydrant as a point of reference, you would say that the dog is in front of the fire hydrant.

If you use the sidewalk as a point of reference, you would say the dog is sitting on the sidewalk.

Either way, the dog is in the same place. You just used a different point of reference to describe it!

▲ What does the waterfront have to do with the fireworks?

▲ Besides using the fire hydrant or sidewalk, how else could you describe this dog's position?

The Bottom Line | An object's position can be described in relation to a point of reference.

Location, Location

Suppose your family is visiting your grandparents in Washington, D.C. "We'll meet you in front of the White House," Grandpa says over the phone. "See you there." *Click.*

"Okay," says Mom, looking at the road signs from the window of the car. "We're at 22nd and H streets. Let's check out a **map** to see where we go from here." You unfold a map of the city like the one below. "Oh, this will be easy to find," you announce. "Most of the city is laid out on a **grid**."

A grid is a pattern of crisscrossed lines, like a checkerboard. Most cities and towns are designed on a grid system in which most streets intersect at right angles. A grid is great for helping you find places because you can easily see where you are, where you need to go, and how to get there.

▼ How would you give directions from the Ellipse to the Lincoln Memorial?

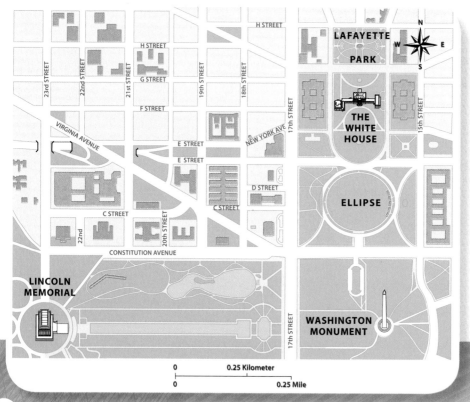

▲ You can't miss this building! But how would you describe its location?

But don't forget your directions—north, south, east, and west. These compass directions help you find places and help you tell others how to get there.

"Here's Lafayette Park," Mom calls out. "Great!" you exclaim. "The White House is just south of there. Let's park the car and walk."

How Far *Is* It?

Sometimes it helps to locate an object or place by using distance. For example, you might locate the White House by describing how far it is from the Lincoln Memorial. You describe distance by using units of length. You may be most familiar with inches, feet, and miles. These units always describe the exact same lengths. An inch is the same length every time it's used. Inches, feet, and miles are part of the English system of measurement.

Another system of measurement is the **metric system**. Centimeters, meters, and kilometers are the metric system's standard units of length. They always describe the exact same lengths—a centimeter is the same length every time it's used. Almost all people in the world use the metric system.

The Bottom Line | You can locate objects and measure distances between them on a map by using the grid and the compass directions.

The Short and the Long of It

Some distances are short. The distance across one red blood cell is about 6/1,000,000 of a meter. That means 10,000 red blood cells, placed side by side, would stretch to a length of only 6 centimeters (2.4 inches)!

Some distances are long. The distance around Earth at the equator is about 40,000 kilometers (25,000 miles). That means if you walked nonstop, day and night, it would take you nearly a year to walk all the way around the equator. The really hard part is that you'd have to walk across water most of the time— since three-quarters of our planet's surface is covered by salty oceans!

▼ **The distance around the world is long . . . compared to your trip to school. How do you think it compares to the distance to the Moon?**

Equator

World Traveler

Each year the Arctic tern travels from the Arctic to Antarctica and back again. That's a round trip of 35,000 kilometers (22,000 miles). Over its 20-year lifetime, an Arctic tern will have traveled a distance equal to flying to the Moon and back!

Movin' on Down

Each September, millions of monarch butterflies start arriving in central Mexico. Thousands of butterflies crowd each fir tree in the mountain forests.

Where do the butterflies come from? How far do they travel to get to Mexico? To answer these questions, scientists started tracking monarch movements. Some even flew in small airplanes along with the butterflies to chart the distance, direction, and time they flew.

▲ Monarch butterflies make the round trip between their summer and winter homes only once in their lifetime.

Each August, people in the northeastern U.S. report seeing millions of butterflies flying southwest. In a few days, butterfly spotters in the Midwest and then in the Southwest report seeing enormous swarms of butterflies flying together.

Scientists collect the reports of the sightings and show them on a map. The map shows the distance and direction the butterflies travel during the late summer and early fall. Those that complete the trip will have traveled about 5,000 kilometers (3,000 miles).

▲ Monarch butterflies start heading south in late summer, traveling up to 160 kilometers (100 miles) per day!

The Bottom Line | An object's motion is described by tracking how its position changes over time.

Fast, Faster, Fastest

Zoom! When a rocket roars into space, its position changes really really quickly. When a sloth crawls along a branch, its position changes really slowly. **Speed** is a measure of how fast or slow something moves. More exactly, speed is how far an object travels in a certain amount of time.

Suppose you know an object's speed. A jackrabbit, or hare, might hop across a field at a speed of 20 meters (66 feet) per second. A tortoise might plod across the field at only 8 centimeters (3.2 inches) per second. How do you find out how far each will travel in a certain time—say, 10 seconds?

◄ **You know which animal is faster. But which covers more distance in a certain time? (Trick question alert!)**

How Far Will It Go?

Let's start with the hare. In 1 second, the hare runs 20 meters (66 feet). So in 2 seconds, it runs 40 meters (131 feet).

$$20 \text{ meters} + 20 \text{ meters} = 40 \text{ meters}$$

And that's the same as multiplying speed by time.

$$2 \text{ seconds} \times \frac{20 \text{ meters}}{1 \text{ second}} = 40 \text{ meters}$$

Look at this table.

Speed (meters/second)	Time (seconds)	Distance (meters)
20	1	20
20	2	40
20	3	60
20	5	100
20	10	200

See the pattern? When you multiply speed by time, you get the distance traveled. Do the math and see for yourself.

$$\text{Distance} = \text{speed} \times \text{time}$$

The hare runs 200 meters (656 feet) in 10 seconds.

Meanwhile, the tortoise moves 8 centimeters (3.2 inches) per second.

$$\frac{8 \text{ centimeters}}{1 \text{ second}} \times 10 \text{ seconds} = 80 \text{ centimeters}$$

The tortoise moves 80 centimeters (32 inches) in 10 seconds.

Of course, you knew which would go farther! The faster an object moves, the more distance it travels in a certain time.

The Wow!

When Speed = Sleep

The world's fastest train is in Shanghai, China. It runs between the airport and downtown Shanghai. The train's speed is about 430 kilometers (267 miles) per hour. This means that a trip that used to take 45 minutes now takes about 8 minutes. If you're a commuter, you get to sleep in about a half hour longer.

▲ **Will these runners have the information they need to figure out their speed?**

Dividing for Speed

What happens if you know the distance something traveled but not its speed? Can you figure out the speed?

Let's say you send a friend the video of your winning run at a track meet. You ran 100 meters (328 feet) in 20 seconds. Your friend sends back the video of her own victory. She ran 250 meters (820 feet) in 50 seconds. If you two got together for a race, who would win?

Math to the rescue! You know the distances traveled. You also know the amount of time it took to travel those distances. So you can use that information to calculate each speed!

$$\text{speed} = \frac{\text{distance}}{\text{time}}$$

Speed equals the distance traveled divided by the time it took to travel that distance.

So, who would win?

Stop. Go. Fast. Slow.

What happens if you know the distance something moved, but it moved at many different speeds while it was traveling?

This can happen during an 80-kilometer (50-mile) bike ride. You may streak downhill at 50 kilometers (31 miles) an hour—as fast as a car on a city street. When you ride uphill, however, you'll be lucky to do 8 kilometers (5 miles) an hour. What's more, you may stop once or twice for water or to catch your breath. So at those times, your speed would be zero.

After 4 hours, you are back where you started. How would you describe your speed for the whole bike ride?

You'll need to calculate your **average speed**. To find the average speed, you need to know the total distance, 80 kilometers (50 miles), and the total time

of the ride, 4 hours. Your average speed would be 20 kilometers (12.4 miles) per hour.

$$\text{Average speed} = \frac{\text{total distance}}{\text{total time}}$$

$$= \frac{80 \text{ kilometers}}{4 \text{ hours}}$$

$$= 20 \text{ kilometers per hour}$$

The units for average speed are the same as they are for speed.

$$\text{Average speed} = \frac{\text{units of distance}}{\text{units of time}}$$

▶ Cyclists move at different speeds throughout a race.

How Fast *Is* It?

The headlines report two different speeds. Each presents speed in these terms.

$$\text{Speed} = \frac{\text{units of distance}}{\text{units of time}}$$

Each speed uses different units for distance and for time. In fact, any speed can be written using many different units without changing the speed. The baseball moved at 101 miles an hour. That's the same as 8,888 feet a minute, or 1,778 inches a second! To change units, all you need to do is multiply.

THE DAILY NEWS
"All the news that's fit to print"

PITCHER HURLS BASEBALL AT
101 MILES AN HOUR

THE DAILY NEWS
"All the news that's fit to print"

ROCKET ENTERS
OUTER SPACE AT
672,000 METERS
PER MINUTE

S₀ THAT'S Why!

When is a year *not* a unit of time? When it's a light year! A light year is a unit of distance. It's a humongous distance—like the distance from Earth to stars. To be more precise, a light year is the distance that light travels in a year, or . . .

A light year = $\dfrac{300{,}000 \text{ kilometers}}{1 \text{ second}}$ x $\dfrac{60 \text{ seconds}}{1 \text{ minute}}$ x $\dfrac{60 \text{ minutes}}{1 \text{ hour}}$ x $\dfrac{24 \text{ hours}}{1 \text{ day}}$ x $\dfrac{365 \text{ days}}{1 \text{ year}}$

= 9,460,800,000,000 kilometers per year

Now that's truly humongous!

Slow Movers

At the other end of the speed scale are some of the slowest movers on Earth. A garden snail ekes by at a top speed of 79 centimeters (31 inches) per minute. That means it would take about 48 seconds for a garden snail to move across your desk.

The continents of Australia and Antarctica are moving apart at a speed of about 6 centimeters (2.4 inches) each year. That's about the speed at which your fingernails grow!

Some Different Speeds	
Object	**Speed**
Speed of light	300,000,000 meters per second (186,000 miles per second)
Jet	966 meters per second (0.6 miles per second)
Speed of Earth as it travels around the Sun	108,000 kilometers per hour (67,000 miles per hour)
Cheetah	113 kilometers per hour (70 miles per hour)
Traffic on a city street	56 kilometers per hour (35 miles per hour)
Average person walking	4.8 kilometers per hour (3 miles per hour)
Garden snail	79 centimeters per minute (31 inches per minute)
Seafloor spreading along the Mid-Atlantic Ridge	2.5 centimeters per year (1 inch per year)

▲ This table shows some of the fastest and slowest speeds on Earth.

The Bottom Line

The units for speed are always $\dfrac{\text{units of distance}}{\text{units of time}}$

Every year hundreds of tornadoes tear through the American Midwest. How can meteorologists tell where a tornado will go and when it will get there? Radar and other equipment help, but nothing beats "eyes on the ground." That's where storm spotters come in.

Collecting Data

Storm spotters trained by the National Weather Service collect data during stormy weather. Spotters report any tornado, funnel cloud, large hail, or flash flooding they see. They also report winds greater than 80 kilometers (50 miles) per hour. Storm spotting can be exciting. Collecting weather data while a storm is heading toward you is like taking notes about an elephant as it charges you!

Meteorologists analyze, or study, the data to determine if a danger exists. If so, they warn the public. Meteorologists and storm spotters work together to protect people from extreme weather on the move, including the most feared air motion of all—tornadoes.

Interpreting Data

Why does one storm produce a tornado while another storm does not? To find out, meteorologists study data from both recent and past tornadoes.

One of the most studied tornadoes happened way back on March 18, 1925. Called the Tri-State Tornado, it tore through parts of Missouri, Illinois, and Indiana. It was the fastest, and deadliest, tornado in U.S. history. The numbers in the table show how far the tornado had traveled at various times during its $3\frac{1}{2}$ hours on the ground.

Total Time (Hours)	Total Distance (Kilometers)
0.25	37
1.00	109
2.00	197
3.00	286
3.50	352

Scientists often use line graphs to help them interpret data. Create a line graph with time on the X-axis and distance on the Y-axis. Mark off every 25 kilometers and every quarter hour. Plot the ordered pairs—time and distance—on the graph. For example, the first point is 0.25, 37 and the second point is 1.00, 109. Then connect the points with a line.

Your turn! Now use your graph to answer these questions.

1. The Tri-State Tornado moved 37 kilometers between 1:00 P.M. and 1:15 P.M. What was its speed in kilometers per hour? How do you know?

2. Explain how to use the line graph to find the distance the tornado traveled during its first 1.5 hours on the ground.

3. What was the average speed of the tornado? Show how you figured it out.

4. Suppose that after the tornado had been moving for an hour, it stopped and spun in place for 15 minutes. What would the graph look like?

Chapter 3: Velocity

Speed Is Not Enough

The direction something is moving can be more important than its speed. Just ask one of these **migrating** wildebeests—not that they'd answer.

Each spring, more than a million wildebeests and a half million zebras leave the Serengeti plains of Tanzania, Africa, and trek north. Why the big move? The rainy season is over, and all the grass has been gobbled up. So the great herds sweep northward, leaving dry regions for grazing in lush grasslands. But not to worry—they'll be back. By late fall, the herds head south, following the rains and growing grass.

▲ Suppose you know the speed of these migrating zebras and wildebeests. What else do you need to know to describe their movement?

Getting Clear Directions

The speed of a wildebeest trudging north may be the same as its speed marching south, but its **velocity** is different. Why? Because its direction is different. Velocity is speed in a certain direction.

From the plains of Africa to the fields of your hometown, velocity—speed and direction—is important. Picture a hot, muggy afternoon. You and your friends are playing soccer. Suddenly a teammate on the bench yells, "A tornado just touched down in Centerville. It's moving at 48 kilometers (30 miles) an hour!" "Which way is it moving?" you ask. "Is it heading toward us?"

▲ A tornado often changes its direction, changing its velocity.

Tight Turns

Jet pilots flying at air shows must be especially careful about their speed and direction. Four to six jets flying in formation turn, roll, and loop at speeds of up to 800 kilometers (497 miles) per hour. Flying in the right direction is really, really important—the planes are often only 50 *centimeters* (20 inches) apart!

The Bottom Line | Velocity is speed in a certain direction.

21

Ch-Ch-Ch-Changes

You don't have to be concerned with a tornado to experience velocity. Any object that moves at any speed in a specific direction has velocity. You have a certain velocity as you shuffle to the bathroom in the morning. A train has a different velocity as it races down the track toward its destination.

What happens if an object's speed changes? Or how about if its direction changes? What does that do to the object's velocity?

If the goalie on a soccer team blocks a shot away from the goal, the ball's direction changes. This means the ball's velocity changes, too.

If an elevator slows as it reaches the top floor, its direction is the same, but its speed is lower. The elevator's velocity changes.

A cheetah speeds up as it's about to run down its prey. The cheetah's direction is the same, but its speed changes. And— you guessed it—the cheetah's velocity changes, too.

▼ **When a ball changes direction, its velocity changes.**

The Thrill of Changing Velocity

AAAEEEEE! One of the world's favorite rides is a roller coaster. The entire ride depends on changing velocity! First, the roller coaster pauses at the top of the loop. It speeds up as it coasts downhill, moving fastest at the bottom of the loop. Then the coaster slows as it climbs uphill. Throughout the ride, the coaster's speed and direction change constantly—so its velocity changes constantly, too. What a thrilling ride!

What else depends on changing velocity? Just about anything that moves. A rocket blasting into space changes velocity. Blood flowing from your heart to your feet changes velocity. A dog running to catch a stick changes velocity.

So how do you get velocity to change? What do you need to do to get something to speed up, to slow down, or to change direction?

▶ The roller coaster cars constantly change velocity as they loop the loop.

Velocity changes if speed changes or if direction changes.

A Push for Change

You already know how to change an object's velocity. You need a **force**—a push or a pull. There is no **change in velocity** unless there is a force. No exceptions, ever. None.

How do these basketball players change velocity up and down the court? One way is for a player to push her wheels with more force. Then she moves faster. How about direction? Pushing on one wheel while pulling on the other changes that.

The Wow!

Whoosh—What a Force

If you've ever felt the wind blow in your face, you know it's a force to be reckoned with. It can be put to use, too. Wind, or moving air, pushes the blades of wind turbines. The blades change velocity as they turn. This motion generates electricity that travels through wires to homes and businesses.

◀ The players change their own velocities and the velocity of the ball to score.

It's All the Same

You may be thinking, "I don't have to use force to stop my skateboard. Even if it doesn't hit a wall, it stops by itself."

Don't believe everything you see! If the skateboard slows or stops, there must be a force acting on it. The force you are not seeing is **friction**. Friction is a force that works to stop motion.

Reduce friction by oiling the wheels or smoothing the street, and the skateboard would move farther before slowing down.

Out in space there is no air. That means there is no friction. Push the skateboard in outer space, and it will travel in a straight line with the speed you gave it forever—that is, unless it crashes into an asteroid or a planet.

If you found a way to remove friction on Earth completely, you would get the same results. Oh, and if you did find a way, you'd be famous! Now that's poetry in motion.

▶ The less friction acts on the skateboard, the farther the board will move.

The Bottom Line | An object's velocity will not change unless a force acts on it.

HOW DO WE KNOW?

Life on the Line

Bluefin tuna are among the fastest fish. They can briefly hit speeds of 95 kilometers (59 miles) per hour when chasing herring, squid, anchovies, and other favorite foods. Unfortunately, bluefin tuna are one of *our* favorite foods. They are no match for fishing boats.

Every day, fishing fleets around the globe catch thousands of bluefin. They satisfy the world's appetite for a raw meal called sushi. Nibble by nibble, one of the ocean's giant fish is steadily disappearing.

Nowhere to turn . . .

Biologists worry that bluefin could face extinction. In the 1970s, there were twice as many bluefin swimming in the eastern Atlantic Ocean as today. In the western Atlantic, there were ten times more. How can we protect the tuna from overfishing?

The problem is that bluefin, unlike other animal species, know no boundaries. A pride of lions might stick to a large territory of African grasslands. That can make it easier to protect the big cats from being hunted. But bluefin tuna roam from ocean to ocean, hidden beneath the waves as they crisscross the world.

Barbara Block

Marine
Biologist

STANFORD
UNIVERSITY

Everybody has a favorite athlete. Barbara Block's is a fish. "They can do things I marvel at," she says. Barbara was a young girl when she saw her first bluefin hanging from a fisherman's scale on Cape Cod, Massachusetts. The big, blue fish looked sleek and powerful.

Later, in college, Barbara learned just how powerful bluefin are. They swim faster and farther than most other ocean creatures. But where do they go? "Until recently, we didn't know," Barbara says. Electronic tags now let biologists follow tuna.

Barbara spends two to three months a year at sea tagging bluefin. She has learned a lot. The tuna follow the same routes as they crisscross the oceans. The tags also show they return each year to mate in places like the Gulf of Mexico and Mediterranean Sea.

Barbara thinks banning tuna fishing in those places could help protect the bluefin from being hunted to extinction. If only, Barbara says, we learned to appreciate the bluefin and their dark-red muscles as a marvel of strength and speed—and not just a delicious dish.

▲ Barbara and her tagging team jump into action seconds after landing a tuna. Everyone plays a role so the tuna is tagged—and freed—quickly.

"Fish on!" The boat crew snags another bluefin tuna. This one is as long as a couch. As the fish rises, it comes face to face with Barbara Block—and six biology students. "One, two, three—pull!" They heave the 135-kilogram (298-pound) fish from the ocean. Now they have just minutes. "Irrigation!" A student runs seawater over the fish's gills. The water supplies vital oxygen. Another student covers the tuna's eyes to calm it. Meanwhile, Barbara delicately sews in an electronic tracking tag. The crew then returns the fish to the ocean. With a flick of the tail, it's gone. Silently, the tag records every move the tuna makes. Barbara wonders, "Will I ever see it again?"

Electronic tags let biologists follow bluefin tuna as they dart around the ocean. Archival tags, like the one to the right, are implanted in tuna. The tuna must be caught again to recover the tag and its data. Other tags, called satellite tags, pop off after a few months and float to the surface. There, the tags transmit the data into space to satellites circling Earth. Researchers then download the information. Both types of tag record where and when the tuna swam, as well as how deep and fast.

This flexible, 25-centimeter (9.8-inch) stalk sticks into the water. It holds instruments that measure water temperature and record light levels, which help pinpoint where the tuna has been.

The battery-powered tag's body has a clock and instruments that measure the tuna's temperature and how deep it dives—information that reveals when it has fed.

MATH CONNECTION

Go the Distance!

An electronic tag showed that a young bluefin tuna crossed the Pacific Ocean three times—across, back, and across again—in just 600 days. Each one-way crossing is 12,000 kilometers (7,500 miles).

> What's the total distance, in kilometers, the bluefin swam in 600 days? What's the total distance in miles?

> What's the average distance, in kilometers, the bluefin swam each day? What's the average distance in miles?

Oh, By the Way

Atlantic bluefin can be as long as a car—that's 4 meters (13 feet) long!

▲ After only 3 minutes out of water, the bluefin is helped back into the ocean.

From slowly crawling sloths to fast-flying rockets—you've learned a lot about motion. Here's your chance to show what you know about a world on the move. On a sheet of paper, do the activities and answer these questions.

1. In the story about the tortoise and the hare, the hare knows he can run faster than the tortoise. So the hare takes a nap while the tortoise keeps on walking. How long must the hare nap for the tortoise to win the race? The field is 100 meters (328 feet) long. (pages 13 and 15)

2. A bike rider takes 2 minutes to travel a track that is 1,341 meters (4,400 feet) long. What is her speed? (page 14)

3. A parachute ride begins with a speed of zero. The chutes take riders 76 meters (249 feet) up in 40 seconds. The ride ends with the chutes falling 76 meters (249 feet) in 20 seconds. What is the average speed for each half of the ride? What is the velocity of the first half of the ride? (pages 15 and 20)

4. On a sheet of paper, draw a half-pipe. Draw three skateboarders at different positions on the half-pipe. At each position, tell if the skater's velocity is changing. If it is changing, tell how. (page 20)

▲ **What happens to this skateboarder's velocity as he soars above the half-pipe and then heads back down? (pages 22 and 23)**

Glossary

average speed **(n.)** the total distance an object travels divided by the total time it takes to go that distance (p. 15)

$$\text{Average speed} = \frac{\text{total distance}}{\text{total time}}$$

change in velocity **(n.)** a change in an object's speed, direction, or both. The change in speed may be either an increase or decrease. (p. 24)

force **(n.)** a push or pull applied to an object (p. 24)

friction **(n.)** a force that resists the motion of objects or materials that are in contact with each other (p. 25)

grid **(n.)** a set of horizontal and vertical lines that form rows of squares, usually on a map. The squares identify the locations of specific objects or areas. (p. 8)

map **(n.)** a drawing of a town, city, state, country, or the entire Earth that shows the location of features (buildings, streets, parks, lakes, mountains) relative to one another (p. 8)

metric system **(n.)** the standard system of measurement used around the world. It is based on the meter as a unit of length, the kilogram as a unit of mass, and the second as a unit of time. The metric system is a decimal system. Every unit of length can be changed to a larger unit by multiplying by 10, 100, 1,000, and so on. Smaller units are found by multiplying by 0.1, 0.01, 0.001, and so on. (p. 9)

migration **(n.)** the seasonal movement of certain animals, mostly birds and fish, to distant places for breeding or feeding (p. 20)

molecule **(n.)** a group of two or more atoms held together by chemical bonds (p. 5)

motion **(n.)** a change in position or place (p. 5)

point of reference **(n.)** a point or location to which other points or objects are referenced or compared (p. 6)

position **(n.)** a place or location (p. 6)

speed **(n.)** the distance an object travels during a certain time (p. 12)

$$\text{Speed} = \frac{\text{distance}}{\text{time}}$$

velocity **(n.)** the speed and direction of an object's motion (p. 21)

Index

About the Author Glen Phelan's fascination with science was sparked when he was a teenager by the lunar missions of the Apollo Program. He shares his fascination through teaching and writing. Learn more at www.sallyridescience.com.